SMELLING

by Robin Nelson

first step nonfiction

Lerner Publications Company · Minneapolis

Smelling is one of my
senses.

I smell with my nose.

I smell something good.
I smell flowers.

I smell apples.

I smell something bad.
I smell a wet dog.

I smell fish.

I smell something nice.
I smell **soap.**

I smell warm pie.

I smell something not so nice. I smell a **skunk.**

I smell a dump.

I smell something yummy.
I smell cake.

I smell popcorn.

I smell something yucky.
I smell a **pigpen.**

I smell stinky feet.

I smell many things.

What do you smell?

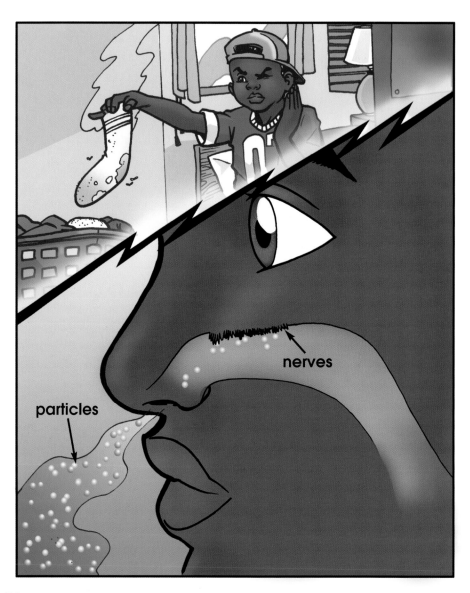

How do you smell?

Smells are made of tiny particles. These particles float in the air. When you smell something, the particles float into your nose. The particles stick to the mucus in your nose. There are special nerves in your nose under the mucus. The nerves send a message to your brain. Your brain tells you what you smell.

Smelling Facts

 Many animals, such as dogs and deer, have a better sense of smell than humans do.

 A dog's sense of smell is 100 times better than a human's.

 People who cannot smell have a condition called anosmia.

 If your nose is at its best, you can identify between 4,000 to 10,000 smells!

As you get older, your sense of smell gets weaker. Children have a better sense of smell than their parents and grandparents.

 A shark can smell its prey with as few as five drops of blood in the water.

 Scientists have identified about 17,000 different kinds of smells.

Glossary

 pigpen – the place where pigs live

 senses – the five ways our bodies get information. The five senses are hearing, seeing, smelling, tasting, and touching.

 skunk – an animal that can make a strong smell to make other animals go away

 soap – helps take dirt off of things when it is mixed with water

Index

Cover image used courtesy of: Corbis Royalty Free Images.

Photos reproduced with the permission of: © 2001 Nick Dolding/Stone, pp. 2, 22 (second from top); Corbis Royalty Free Images, p. 3, 4, 15; © Richard Cummins, p. 5; © Gail Mooney/ CORBIS, p. 6; © 2001 White.Packert/The Image Bank, p. 7; © Myrleen Cate/Photo Network, pp. 8, 22 (bottom); © 2001 PhotoDisc, pp. 9, 13, 16, 17; © W. Perry Conway/CORBIS, pp. 10, 22 (second from bottom); © Betty Crowell, pp. 11, 14, 22 (top); Brand X Pictures, p. 12.

Illustration on page 18 by Tim Seeley.

Lerner Publications Company
A division of Lerner Publishing Group
241 First Avenue North
Minneapolis, MN 55401 U.S.A.

Website address: www.lernerbooks.com

Library of Congress Cataloging-in-Publication Data

Nelson, Robin, 1971–
 Smelling / by Robin Nelson.
  p. cm. — (First step nonfiction)
 Includes index.
 Summary: An introduction to the sense of smell and the different things that you can smell.
 ISBN: 0–8225–1263–7 (lib. bdg. : alk. paper)
 1. Smell—Juvenile literature. [1. Smell. 2. Nose. 3. Senses and sensation.] I. Title.
II. Series.
QP458 .N45 2002
612.8'6—dc21   2001003964

Manufactured in the United States of America
2 3 4 5 6 7 – DP – 08 07 06 05 04 03